# YOUR KNOWLEDGE HAS VALUE

- We will publish your bachelor's and master's thesis, essays and papers

- Your own eBook and book - sold worldwide in all relevant shops

- Earn money with each sale

Upload your text at www.GRIN.com
and publish for free

**Bibliographic information published by the German National Library:**

The German National Library lists this publication in the National Bibliography; detailed bibliographic data are available on the Internet at http://dnb.dnb.de .

This book is copyright material and must not be copied, reproduced, transferred, distributed, leased, licensed or publicly performed or used in any way except as specifically permitted in writing by the publishers, as allowed under the terms and conditions under which it was purchased or as strictly permitted by applicable copyright law. Any unauthorized distribution or use of this text may be a direct infringement of the author s and publisher s rights and those responsible may be liable in law accordingly.

**Imprint:**

Copyright © 2018 GRIN Verlag
Print and binding: Books on Demand GmbH, Norderstedt Germany
ISBN: 9783668693234

**This book at GRIN:**

https://www.grin.com/document/423822

Patrick Kimuyu

# Application of Nanotechnology in Improving the Biofuel Generation

GRIN Verlag

**GRIN - Your knowledge has value**

Since its foundation in 1998, GRIN has specialized in publishing academic texts by students, college teachers and other academics as e-book and printed book. The website www.grin.com is an ideal platform for presenting term papers, final papers, scientific essays, dissertations and specialist books.

**Visit us on the internet:**

http://www.grin.com/

http://www.facebook.com/grincom

http://www.twitter.com/grin_com

# Application of Nanotechnology in Improving Bioenergy Generation

Patrick Kimuyu

## Contents

Introduction ............................................................................................................................. 2
Nanoscale Materials ................................................................................................................ 2
    Nanobiocatalysis ................................................................................................................. 3
Nanotechnology for Biofuel Production from Butchery Waste .............................................. 4
Nanostructured Materials for Bioelectrochemical Systems .................................................... 5
Nanofarming Technology for Obtaining Biofuel from Algal Biomass ................................... 6
Conclusion .............................................................................................................................. 6
References ............................................................................................................................... 7

## Introduction

For centuries, energy crisis has been a major challenge that has affected economic development in almost all nations around the globe. As a result, so as to meet the severity of this day-to-day requirement, different nations have shifted their focus towards alternative energy sources that contain diverse energetic potential, and have the ability of meeting the economic concerns among nations. Bioenergy is one of the main non-conventional resources of energy that is comprised of energy obtained from life forms (Malik, & Sangwan, 2012). High dependence on conventional sources of energy that have been directly or even indirectly obtained from fossil fuels, for example, petroleum and coal have demonstrated detrimental effects on the environment in different nations worldwide. Bioresources are eco-friendly because they are not created, are self-prevalent and everywhere, not to mention that they do not leave toxic end products after processing. Some of the major bioresources include biomass, carcasses, dead plants and cow-dung. However, there is limited biomass utilization in energetic application due to inadequate biochemical modification techniques that may facilitate profitability. Nanotechnology refers to a new dimension of material sciences that targets entities miniaturization so as to achieve improved performance through functionalities improvement. It is a dimension that is multidisciplinary in nature since it originates from diverse, overlapping natural sciences fields. It is a technology that has provided faster and reliable methods of optimizing energy generation from the bio-sources. Therefore, the main purpose of this paper is to explore the use of biomass and nanotechnology as alternative energy source.

## Nanoscale Materials

Nanoscale materials are those materials that are products of nanotechnological principles along with their procedures, and they are developed through two main methodologies. The first methodology is termed as top-down technique whereby it involves the creation of smaller entities from the larger entities while the second one is called bottom-up technique (creation of larger entities from smaller objects (Malik & Sangwan, 2012). Several methods have been established to design nanomaterials, and can be categorized as physical, biological or chemical processes. Consequently, nanoparticles highly contribute in bringing about microbial species biotransformation, with the aim of maximizing bio-products generation that solicits biofuel as a principal component (Ramsurn & Gupta, 2013). Significantly, this is one way of meeting the challenge of energy predicament since it is cost effective, easy and reliable. Furthermore, metal nanoparticles, especially the ones geared for energy production have been used as nano-catalysts, nanostructured amorphous alloys and nanoclusters, metals and their oxides in chemical reactions to improve efficiency (Malik & Sangwan, 2012).

## Nanobiocatalysis

Nanobiocatalysis simply refers to the ways for efficient as well as economic processing of biomass. Evidently, several industries have engaged in the oxidation of saturated hydrocarbons such as cyclohexane family so as to meet their energy needs (Serrano, Rus & Garcia-Martinez, 2009). However, the methods available for these industries in this process have been noted to be energetically unfavorable since they require high temperatures and pressure, with this problem elevating when living systems are used as they use room temperature (Malik & Sangwan, 2012). Direct use of oxygen in the oxidation of cyclohexanol is an endothermic process, with the main purpose of this process being the search for alternative oxidizers that aid in prevailing over the reaction endorgenicity, and saves considerable energy amount. It is an activity that has led to the integration of different alternative oxidizers for this process, with nano-catalyst, thus; providing a momentous breakthrough. According to different studies, it has been affirmed that nanophased catalyst particles such as iron, nickel or cobalt, are highly potent energy savers in their product yields (Malik & Sangwan, 2012). In this process, effective control of surface functionalities helps in catalyzing the highly diverse as well as non-uniform bio-sources, for example, plant biomass, animal fats and cellular remains alongside processed compounds chemical modification like alcohols. Malik and Sangwan (2012) claim that the process of surface engineering in such catalysts facilitates the modification of biochemical features of their active sites that makes them increasingly robust.

In order to achieve a sustainable and reliable biomass utilization to meet energy requirements, there are complex and even consistent concerns that have to be satisfied on a large scale (Sekhon, 2014). One of the main requirements encompasses the application of efficient pre-processing techniques as well as degrading-enzymes development in a readily and efficient way (Malik & Sangwan, 2012). In this sense, there must be a degrading enzyme, for example, the hydrolases, which has to be accompanied by stringent conditions of processing that in turn requires the use of strong acid treatments under high temperatures. Organic solvents, for instance, acetone have to be used for conversion of complex feedstock compounds into readily and energetically useful intermediates through the help of microorganisms (Malik & Sangwan, 2012).

The main problem that affects the efficiency of biocatalysis each time is the hot area of concern that affects this technology, and in order to mitigate this problem, enzyme immobilization has been broadly explored and adapted. Nanotechnology has been observed as an airlifter in this case, with the nanoparticles continuing to produce better enzyme immobilization (Ramsurn & Gupta, 2013). It is a breakthrough that has distinct advantages, for example; the entire process becomes far much easy and consumes less time, not to mention that it has the benefit of reusability of highly expensive enzymes through their recovery (Malik & Sangwan, 2012). In addition, the significance of exploring an energy source that is everlasting as well as crisis-mediating persists. For example, enzyme-free cellulose will be used for catalytic conversion of the complex carbohydrate cellulose. Conversely, since it suffers limitations due to low specific activity, inactivation susceptibility and high recovery difficulties, it has been immobilized so as to optimize its performance under high temperatures and pH conditions that are needed for the conversion of cellulose (Malik & Sangwan, 2012).

Regrettably, scholars have noted that immobilization through several conventional methods, causes significant loss of the concerned enzyme's specific activity, with nanotechnology assuming a potent weapon-like nature in this case. Immobilization may be battered upon through incorporating nanoparticles, which have high surface area to volume ratios, as this makes the process economically practical and simpler (Malik & Sangwan, 2012; Serrano, Rus & Garcia-Martinez, 2009). Thus, nanotechnology has provided a clarification on improved enzyme loading, biocatalyst recovery along with its reusability, although it requires much impetus to ensure an incessant operation. Recent interventions concerning enzyme aggregate coatings where the problem of enzyme loading was because of the inability to triumph over the monolayer in nanostructured medium has been determined (Malik & Sangwan, 2012). It is a technique that involves several processes and has not only led to increased enzyme loading, but has also increased enzyme activity and assured long-term enzyme stability (Rajvanshi, 2008).

## Nanotechnology for Biofuel Production from Butchery Waste

Large animal fats have been discarded in many parts of the world, with people being ignorant of the benefit they can get from these animal fats (OECD, 2013). Fortunately, these animal fats can be reduced chemically through a process called trans-esterification into energetically useful materials, for instance, biofuels (Malik & Sangwan, 2012). Biodiesel, for example can be used in place of the increased use of petroleum that has detrimental effects on the environment. Notably, biodiesel is chemically extracted from trans-esterification of animal fats, vegetable oils among other waste substances that contain animal fats texture in nature. Trnas-esterification is the best strategy since it is suitable for bringing down animal fats' viscosity, and the addition of catalyst based nanoparticles such as cobalt, leads to improved biofuel yields. Therefore, the idea of harnessing biofuels in extraordinary quantities, under the help of nanoscaled catalyst particles has been observed to highlight the unseen potential of nanotechnology in bringing a lifestyle revolutionary change and meet energy requirements in a more convincing way (Malik & Sangwan, 2012).

## Nanostructured Materials for Bioelectrochemical Systems

Several scholars and scientists have also exploited microbial systems in the generation of electrical energy through nanomaterials intervention. They have employed nanomaterials in their natural and even their engineered forms so as to mitigate energy crisis problem alongside tapping the infinite potential of the bioenergy found in these microorganisms (Malik & Sangwan, 2012). Evidently, nanomaterials have facilitated electric current generation in microbial bioelectrochemical systems; systems that are similar to the conventional cells. Remarkably, these systems are categorized into two main sub-groups, namely, microbial fuel cell (MFCs) along with microbial electrolysis cells (MECs). MFCs are responsible for the production of electricity while the MFCs aid in the production of hydrogen through electrolysis process (Malik & Sangwan, 2012).

In the aforementioned cases, energy comes from the oxidative metabolism of the electrochemically active microorganisms (bacterial species) that are responsible for catalyzation of the electron generation from inorganic substances such as acetate, waste water or starch sources. Then, they utilize this process to generate energy through electrical circuit completion (Malik & Sangwan, 2012). Nanoporous membranes have, however, demonstrated a high degree of being used as low cost, and have selective membranes of proton transfer compared to the ones that are used currently. With respect to these biochemical interventions, these species of bacteria are used as anodic material, with several species being optimized to function as nanowires to help in the extracellular electrons transfer.

According to recent research, it has been affirmed that electron transfer proteins play significant roles in the making of the cytochromes at the surface of the membrane. Scholars have found that these proteins remain actively involved in the synthesis of nanowire and that they facilitate electron transfer among diverse species of microbes (Malik & Sangwan, 2012). Moreover, these bio electromagnetism traits have been enhanced to a better scale through incorporation of genetically modified viruses, taking the form of assemblers of the engaged nanowires. As a result, incorporation of nanostructured materials in this approached has shown benefits because of their rich attributes of high surface are that facilitates high electron fluxes that has led to overall cost reduction as well as reposing scientific trust in evolving this unexplored energy resource.

## Nanofarming Technology for Obtaining Biofuel from Algal Biomass

Notably, different nations have turned their focus of bioenergy industries on energy that is obtained from biofuels that are based on alcohol. The most favored and optimized bioenergy sources in biofuels forms are natural and synthetic fermentation products such as ethanol, pentanol and isobutanol (Malik & Sangwan, 2012). The key factor that influences biofuel harnessing according to this approach is the processing of short-chained aliphatic alcohols, a complex activity because of their inherent cytotoxicity. Although it may be faced by different challenges during and after the processing, people have widely anticipated algal biomass as the next remedy for energy requirements (OECD, 2013). One of the main setbacks of the current biofuels extraction technologies from algae is that algae are killed after the extraction process is over (Malik & Sangwan, 2012). Fortunately, efficiency of biocatalysts, and high selection of mesoporous membranes that are supplemented with nanoparticle adsorption are the basis of yield production and they help in eliminating the problem of killing algal biomass. The popular method used in biofuels harnessing in this case is the removal of water from fractional distillation ethanol products.

## Conclusion

Nanotechnology has proved to be highly significant in solving some of the global critical problems, in the aspects of energy, manpower requirement, faster and more reliable productivity. For the past two decades, nanotechnology has been termed as a major tool for energy crisis alleviation in the world. It is a significant technology that has provided a great advantage to the daily mankind problems that are correlated to the sheer energy requirement.

Although a number of studies have confirmed the benefits of this technology in the energy sector, it is unfortunate that it has not been adequately accepted especially in the developing economies. With regard to this, although nanotechnology may be deemed to be the remedy for energy crisis in the world, commercial application concerns have to be addressed because nanomaterials may pose some critical perils to human health and environment due to their novel properties and small size.

## References

Malik, P., & Sangwan, A. (2012). Nanotechnology: A tool for Improving Efficiency of Bio-Energy. *Journal of Engineering, Computers & Applied Sciences, 1*(1), 37-47.

OECD (2013). Nanotechnology for Green Innovation, OECD Science. *Technology and Industry Policy Papers, 3*(5), 7-32. Retrieved from http://www.oecd-ilibrary.org/docserver/download/5k450q9j8p8q.pdf?expires=1412423182&id=id&accname=guest&checksum=5ECCBFAC1D5D312726B1199DEEBD9118

Rajvanshi, A. (2008). Langmuir Approach to Rural Development. *Current Science, 95*(7), 2-7. Retrieved from http://www.nariphaltan.org/langmuirrural.pdf

Ramsurn, H., & Gupta, R. (2013). Nanotechnology in Solar and Biofuels. *ACS Sustainable Chemistry & Engineering, 1*(7), 779-797. Retrieved from http://pubs.acs.org/doi/abs/10.1021/sc400046y

Sekhon, B. (2014). Nanotechnology in agri-food production: an overview: *Nanotechnology Science and Applications, 7*(2014), 31-53.

Serrano, E., Rus, G., & Garcia-Martinez, J. (2009). Nanotechnology for sustainable energy. *Renewable and Sustainable Energy Reviews, 13*, 2373-2384. Retrieved from http://www.ugr.es/~grus/publications/SerranoRusGarcia09_RenewSustEnergRev.pdf

# YOUR KNOWLEDGE HAS VALUE

- We will publish your bachelor's and master's thesis, essays and papers

- Your own eBook and book - sold worldwide in all relevant shops

- Earn money with each sale

Upload your text at www.GRIN.com and publish for free